Math Mammoth
Grade 6 Tests and
Cumulative Reviews

for the complete curriculum
(Light Blue Series)

Includes consumable student copies of:

- Chapter Tests
- End-of-year Test
- Cumulative Reviews

By Maria Miller

Contents

Grade 6, Chapter 1

End-of-Chapter Test

Instructions to the student:

A calculator is not allowed. Answer each question in the space provided.

Instructions to the teacher:

My suggestion for grading is as follows. The total is 23 points. You can give partial points for partial solutions. Divide the student's score by 23 and multiply by 100 to get a percent score. For example, if the student scores 17, divide $17 \div 23$ with a calculator to get 0.7391. The percent score is 73.9% or 74%.

Question #	Max. points	Student score
1	2 points	
2	2 points	
3	2 points	
4	4 points	
5	4 points	

Question #	Max. points	Student score
6	2 points	
7	2 points	
8	2 points	
9	3 points	
TOTAL	**23 points**	/ 23

Chapter 1 Test

1. Divide 758,908 ÷ 72 and indicate the remainder, if any. Use long division.

2. Solve 21 ÷ 11. Round to three decimal digits.

3. A box of 75 flashlights costs $937.50. Find the cost of one flashlight.

4. You can scan one page of a book in 43 seconds. Working at the same speed, how long will it take you to scan all 234 pages of the book?
   Give your answer in minutes and seconds.

5. Solve.

   a. 3^3

   b. 1^{10}

   c. 50^2

   d. 10^5

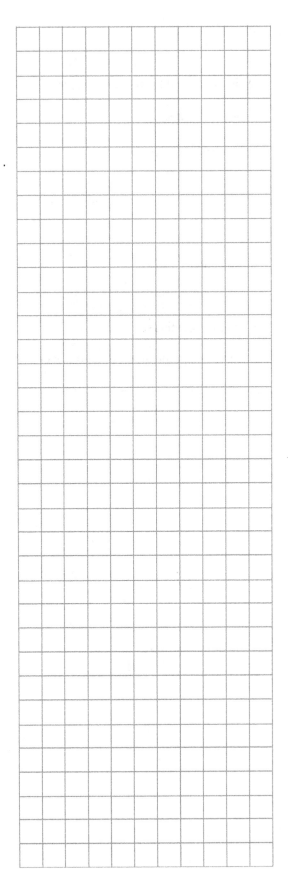

6. The perimeter of a square is 56 cm. What is its area?

7. Write in normal form (as a number).

   a. $5 \times 10^8 + 4 \times 10^6 + 3 \times 10^5$

   b. $1 \times 10^9 + 6 \times 10^8 + 2 \times 10^4 + 1 \times 10^2$

8. Write in expanded form, using exponents (as in the original in #7).

   a. 560,000

   b. 9,108,000

9. Round the numbers.

   a. 2,998,601 to the nearest ten thousand

   b. 483,381,902 to the nearest ten million

   c. 19,993,740 to the nearest million

Grade 6, Chapter 2

End-of-Chapter Test

Instructions to the student:

A calculator is not allowed. Answer each question in the space provided.

Instructions to the teacher:

My suggestion for grading is as follows. The total is 30 points. You can give partial points for partial solutions.

Divide the student score by 30 and multiply by 100 to get a percent score. For example, if the student scores 25, divide 25 ÷ 30 with a calculator, getting 0.833333.... The percent score is 83%.

Question #	Max. points	Student score
1	3 points	
2	3 points	
3	4 points	
4	2 points	
5	4 points	
6	2 points	

Question #	Max. points	Student score
7	3 points	
8	2 points	
9	2 points	
10	2 points	
11	3 points	
TOTAL	30 points	/30

Test - Chapter 2

1. Write an expression.

    a. the quotient of x squared and 7

    b. the quantity 5 minus y, cubed

    c. 3 times the quantity $2s$ minus 5

2. Find the value of these expressions.

a. $(100 - 80) \cdot 2 - 20$	**b.** $480 \div 2^3$	**c.** $32 + 0^5 \cdot 12 \div 4$

3. Find the value of the expressions.

a. $2x + 10$ when $x = 5$	**b.** $x^2 + 10$ when $x = 5$
c. $\dfrac{40 - x}{5}$ when $x = 5$	**d.** $40 - \dfrac{5}{x} =$ when $x = 5$

4. Write an expression.

    You purchased a book for p dollars and three pencil cases for t dollars each. What was the total cost?

5. Simplify the expressions.

a. $a \cdot a \cdot a \cdot a$	**b.** $a + a + a + a$
c. $x \cdot x \cdot 5 \cdot 2$	**d.** $8d - 2d + 7$

6. Multiply using the distributive property.

a. $5(x + 6) =$	**b.** $2(9 + 5y) =$

7. Solve the equations.

a. $6x = 144$	b. $y + 78 = 134$	c. $\dfrac{x}{16} = 3$

8. Write an equation EVEN IF you could easily solve the problem without an equation!
   Then solve the equation.

   The perimeter of a square is 164 units. How long is its side?

9. Plot these inequalities on the number line.

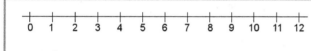

   a. $x \geq 5$

   b. $x < 8$

10. Solve the inequality $x + 3 < 20$ in the set $\{15, 16, 17, 18, 19, 20\}$.

11. Calculate the values of y according to
    the equation $y = 9 - x$.

x	0	1	2	3	4	5	6	7	8	9
y										

   Now, plot the points.

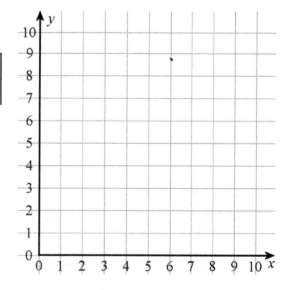

Grade 6, Chapter 3

End-of-Chapter Test

Instructions to the student:
A calculator is not allowed. Answer each question in the space provided.

Instructions to the teacher:
A calculator is not allowed. Since the test is long, consider allowing the student to take a break in between, or administer it in two parts.

My suggestion for grading is as follows. The total is 66 points. You can give partial points for partial solutions.

Divide the student score by 66 points and multiply by 100 percent to get a percent score. For example, if the student scores 51, divide 51 ÷ 66 with a calculator, getting 0.77272727. The percent score is 77%.

Question #	Max. points	Student score
1	6 points	
2	3 points	
3	3 points	
4	4 points	
5	6 points	
6	6 points	
7	4 points	

Question #	Max. points	Student score
8	6 points	
9	3 points	
10	4 points	
11	9 points	
12	2 points	
13	2 points	
14	8 points	
TOTAL	66 points	/ 66

Test - Chapter 3

1. Write as decimals.

a. five thousandths $\quad$ 1.0050	**b.** 382 hundred-thousandths $\quad$ 0.0000382
c. 1 and 3,658 millionths $\quad$ 0.00000 3658	**d.** 94 ten-thousandths $\quad$ 0.00094
e. $\frac{13}{20}$ $\quad$ 0.76	**f.** $8\frac{2}{25}$ $\quad$ 8.8

2. Write as fractions.

a. 2.0045 $\quad \frac{2}{0.45}$	**b.** 0.000912 $\quad \frac{0}{0.0912}$	**c.** 7.49038 $\quad \frac{7}{49.038}$

3. Calculate without a calculator.

a. $0.2 + \frac{5}{1000}$ $\quad \frac{5.2}{1000}$	**b.** $0.07 + \frac{3}{100}$ $\quad \frac{3.07}{100}$	**c.** $2.022 + \frac{33}{1000}$ $\quad \frac{35.022}{1000}$

4. Solve without using a calculator.

    a. 2.31×0.04

    b. $3.38758 \div 7 + 0.045$

5. Round to...

	0.0882717	0.489932	1.299959
the nearest thousandth	0.089	0.481	1.290
the nearest hundred-thousandth	0.08821	0.48903	1.20996

6. Multiply or divide mentally.

a. $0.24 \div 3 =$ 0.8	**b.** $5.4 \div 0.6 =$	**c.** $0.081 \div 0.009 =$
d. $2 \times 0.05 =$ 0.05	**e.** $8 \times 0.009 =$	**f.** $11 \times 0.0005 =$

7. A rectangular plot of land has sides that measure 50.5 m and 27.6 m.
   This plot is then divided into four equal pieces. What is the area of each fourth?

8. Multiply or divide these decimal numbers.

a. $1,000 \times 0.02 =$	**b.** $100 \times 0.0047 =$
c. $10^6 \times 1.097 =$	**d.** $0.6 \div 100 =$
e. $12.45 \div 10,000 =$	**f.** $324 \div 10^5 =$

9. Find the value of the expression $0.04 \div y$, when

a. $y = 4$	**b.** $y = 0.04$	**c.** $y = 10$

10. Change into the basic unit (meter, liter, or gram).

   a. 56 mm   **b.** 9 km

   c. 9 cg   **d.** 16 dl

11. Convert the measurements into the given units.

   a. 2.7 km = _____ m = _____ cm = _____ mm

   b. 5,600 ml = _____ cl = _____ dl = _____ L

   c. 0.6 g = _____ dg = _____ cg = _____ mg

12. A newborn baby weighs 7 pounds 6 ounces.
    Is this more or less than 7.4 pounds?

      no

13. Which is a better deal:
    A 1-pint bottle of honey that costs $7,
    or a 24-oz bottle of honey that costs $12?

    24-oz

14. Divide, giving your answer as a decimal. If necessary, round the answers to three decimal digits.

a. $5.36 \div 0.2$	**b.** $1.6 \div 0.05$
c. $22.9 \div 7$	**d.** $\dfrac{8}{9}$

Grade 6, Chapter 4

End-of-Chapter Test

Instructions to the student:

A calculator is not allowed. Answer each question in the space provided.

Instructions to the teacher:

My suggestion for grading is as follows. The total is 35 points. You can give partial points for partial solutions.

Divide the student score by 35 points and multiply by 100 percent to get a percent score. For example, if the student scores 25, divide $25 \div 35$ with a calculator to get 0.71428. The percent score is 71%.

Question #	Max. points	Student score
1	4 points	
2a	1 point	
2b	2 points	
3	5 points	
4a	1 point	
4b	2 points	
5	2 points	
6a	1 point	
6b	2 points	

Question #	Max. points	Student score
7	3 points	
8a	1 point	
8b	2 points	
8c	2 points	
9	3 points	
10	4 points	
TOTAL	35 points	/ 35

Test - Chapter 4

1. Write the equivalent ratios.

a. $\dfrac{3}{5} = \dfrac{18}{}$	**b.** $2 : 3 = 18 : \underline{}$	**c.** $\underline{}$ to 45 = 2 to 9	**d.** $12 : 30 = \underline{}$ to 5

2. **a.** Draw a picture where there are 4 rectangles for
   each 3 triangles, and a total of 16 rectangles.

   b. Fill in the unit rates:

   _____ rectangles for **1** triangle

   _____ triangles for **1** rectangle

3. Fill in the missing numbers to form equivalent rates.

a. $\dfrac{4 \text{ L}}{10 \text{ m}^2} = \dfrac{}{5 \text{ m}^2} = \dfrac{10 \text{ L}}{ \text{ m}^2}$	**b.** $\dfrac{\$9}{6 \text{ min}} = \dfrac{}{2 \text{ min}} = \dfrac{}{10 \text{ min}} = \dfrac{}{1 \text{ hour}}$

4. A mole can dig 3.6 meters in 36 minutes.

   a. What is the unit rate?

   b. Digging at the same speed, how far can the mole dig in 17 minutes?

5. You can buy 14 song downloads for $2.10.
   How much would 3 songs cost?

6. The length and width of a rectangle are in a ratio of 8:5.
   The shorter side is 15 cm.

   a. Find the longer side of the rectangle.

   b. Find the area of the rectangle.

7. You are mixing juice concentrate and water in a ratio of 1:7.
How much water and how much concentrate do you need
to make 4 liters of diluted juice?

8. A large passenger airplane burns about 35 gallons of fuel per 7 miles.

   a. Write a rate from this, and simplify it to the lowest terms.

   b. How far can the airplane travel with 500 gallons of fuel?

   c. How many gallons will the airplane need to travel 150 miles?

9. Anita and Michael divided a job of folding advertisements for inserts in 1,200 newspapers
in a ratio of 3:5. Calculate how many inserts each one of them folded.

10. Use ratios to convert the measuring units. 1 in. = 2.54 cm, and 1 ft = 30.48 cm.

a. 60 cm into inches
b. 4.5 feet into cm

Grade 6, Chapter 5

End-of-Chapter Test

Instructions to the student:

A calculator is not allowed. Answer each question in the space provided.

Instructions to the teacher:

My suggestion for grading is as follows. The total is 36 points. You can give partial points for partial solutions.

Divide the student score by 36 to get a percent score. For example, if the student scores 24, divide 24 ÷ 36 with a calculator, getting 0.6666666... The percent score is 67%.

Question #	Max. points	Student score
1	6 points	
2	9 points	
3	2 points	
4	2 points	
5	2 points	
6	2 points	
7	2 points	

Question #	Max. points	Student score
8	2 points	
9	2 points	
10	3 points	
11	2 points	
12	2 points	
TOTAL	36 points	/ 36

Test - Chapter 5

A calculator is not allowed.

1. Write as percentages, fractions, and decimals.

a. _____ % = $\dfrac{45}{100}$ = _____	**b.** 179% = $\dfrac{}{}$ = _____	**c.** _____ % = $\dfrac{}{}$ = 0.02

2. Fill in the table. Use mental math.

percentage / number	5,200	80	9
1% of the number			
3% of the number			
70% of the number			

3. Write 4/7 as a percentage. Round your
   answer to the nearest percent.

4. A toy costs $12. It is discounted by 30%.
   What is the new price?

5. A cap costs $7.00. Another cap costs 120% of the price of the first cap.
   How much does the second cap cost?

6. A store got a shipment of 120 T-shirts. Forty percent of them are white.
   How many T-shirts are _not_ white?

7. A store window shows 2 red caps and 8 green caps.
   What percentage of the caps are red?

8. The chess club has 24 members, of which 8 are girls.
   What percentage of the members are boys?

9. Annie is 144 cm tall and Jessie is 160 cm tall.
   What percent of Jessie's height is Annie's height?

10. Which is cheaper, jeans that cost $35 and then are discounted by 10%,
    or jeans that cost $40 and then are discounted by 20%?

    How many dollars cheaper is it?

11. Andrew pays 20% of his salary in taxes. Andrew paid $400 in taxes.
    Find Andrew's salary in dollars.

12. A town has 2,100 senior citizens, which is 15% of the total population of the town.
    Calculate the total population of the town.

Grade 6, Chapter 6

End-of-Chapter Test

Instructions to the student:

A calculator is not allowed. Answer each question in the space provided.

Instructions to the teacher:

My suggestion for grading is as follows. The total is 23 points. You can give partial points for partial solutions. Divide the student score by 23 to get a percent score. For example, if the student scores 17, divide $17 \div 23$ with a calculator, getting 0.739130... The percent score is 74%.

Question #	Max. points	Student score
1	6 points	
2	2 points	
3	2 points	
4	2 points	
5	2 points	

Question #	Max. points	Student score
6	1 point	
7	2 points	
8	4 points	
9	2 points	
TOTAL	23 points	/ 23

Test - Chapter 6

1. Factor the following numbers into their prime factors.

a. 56 / \	**b.** 90 / \	**c.** 101 / \

2. Find the least common multiple of these pairs of numbers.

a. 8 and 6	**b.** 6 and 12

3. Find the greatest common factor of the given number pairs.

a. 98 and 100	**b.** 98 and 35

4. Find four numbers that are multiples of both 6 and 10.

5. Find the LCM of 8 and 10 and the GCF of 8 and 10, and multiply them. What is the product?

6. Which number is a factor of all numbers?

7. Choose two primes between 10 and 30. What is their greatest common factor?

8. First, find the GCF of the numbers. Then factor the expressions using the GCF.

a. The GCF of 24 and 30 is _____

$24 + 30 =$ ____ · ____ + ____ · ____ = ____ (____ + ____)

b. The GCF of 22 and 121 is _____

$22 + 121 =$ ____ (____ + ____)

9. Simplify.

a. $\dfrac{124}{72} =$

b. $\dfrac{65}{105} =$

Grade 6, Chapter 7

End-of-Chapter Test

Instructions to the student:

A calculator is not allowed. Answer each question in the space provided.

Instructions to the teacher:

My suggestion for grading is as follows. The total is 30 points. You can give partial points for partial solutions. Divide the student score by 30 to get a percent score. For example, if the student scores 25, divide 25 ÷ 30 with a calculator, getting 0.833333... The percent score is 83%.

Question #	Max. points	Student score
1	8 points	
2	2 points	
3	2 points	
4	2 points	
5	2 points	

Question #	Max. points	Student score
6	2 points	
7	4 points	
8	3 points	
9	2 points	
10	3 points	
TOTAL	30 points	/30

Test - Chapter 7

A calculator is __not__ allowed.

1. Add or subtract

a. $\dfrac{5}{12} + \dfrac{1}{2} + \dfrac{5}{6}$	**b.** $\dfrac{5}{9} - \dfrac{2}{7}$
c. $2\dfrac{3}{10} + 2\dfrac{11}{12}$	**d.** $7\dfrac{1}{5} - 5\dfrac{7}{15}$

2. The Williams family had 3/4 of a pizza left over. The next day, Joe ate 3/4 of what was left. What part of the original pizza is left now?

3. How many 1/4-kg servings can you get from 5 1/3 kg of meat?

4. Joe had a board that is 4 1/3 ft. long. He cut it into three equal pieces. How long are the pieces, in feet?

5. Multiply. Shade a rectangular area to illustrate the multiplication.

a.  $\dfrac{2}{6} \times \dfrac{2}{3} =$	**b.** 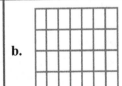 $\dfrac{3}{4} \times \dfrac{5}{7} =$

6. Solve.

a. $\dfrac{6}{7} \div \dfrac{1}{5}$	**b.** $\dfrac{12}{13} \div 2\dfrac{1}{3}$

7. A room measures 11 feet by 8 3/4 feet, and carpeting it costs $2.80 per square foot. Calculate the cost of carpeting the room.

8. Write a real-life situation to match this fraction division. Also, solve it. $2\dfrac{1}{2} \div 3 = ?$

9. How many 1 3/4 ft-pieces can you cut out of a 12-foot piece of string?

10. Mason and Aiden divided a $120 reward in a ratio of 2:3. Then, Aiden gave 3/10 of his money to his dad. How much does Aiden have now?

Grade 6, Chapter 8

End-of-Chapter Test

Instructions to the student:

A calculator is not allowed. Answer each question in the space provided.

Instructions to the teacher:

My suggestion for grading is as follows. The total is 33 points. You can give partial points for partial solutions. Divide the student score by 33 to get a percent score. For example, if the student scores 22, divide 22 ÷ 33 with a calculator, getting 0.666666... The percent score is 67%.

Question #	Max. points	Student score
1	2 points	
2	4 points	
3	8 points	
4	4 points	

Question #	Max. points	Student score
5	9 points	
6	2 points	
7	4 points	
TOTAL	33 points	/ 33

Test - Chapter 8

A calculator is not allowed.

1. Order the numbers 3, −3, −5, and 0 from the smallest to the greatest.

2. Draw a number line jump for each addition or subtraction sentence.

a. −7 + 2 = _____
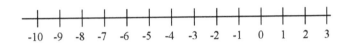

b. −3 + 6 = _____

c. −1 − 5 = _____

d. 2 − 7 = _____

3. Add or subtract.

a.	b.	c.	d.
3 + (−7) = _____	(−1) + (−9) = _____	4 + (−5) = _____	8 − (−2) = _____
(−3) + 7 = _____	1 − 9 = _____	−4 − 5 = _____	−8 − (−2) = _____

4. Use mathematical symbols to express these ideas

    a. the distance of −9 from zero

    b. the opposite of 43

    c. Henry's balance?
    He owes some money, less than $20.

    d. The temperature is colder than −10.

5. Write an addition or subtraction sentence to match the situations.

    a. May owed $3. She borrowed $8 more.
    Now her money situation is _____.

    b. The temperature was 1°C and fell 4°.
    Now the temperature is _____ °C.

    c. A submarine was at the depth of 12 m. Then it rose 5 m.
    It sank 10 m more. Now it is at the depth of _____ m.

6. Plot the points from the equation $y = x - 1$ for the values of x listed in the table.

x	−7	−6	−5	−4	−3	−2	−1	0
y								

x	1	2	3	4	5	6	7	8
y								

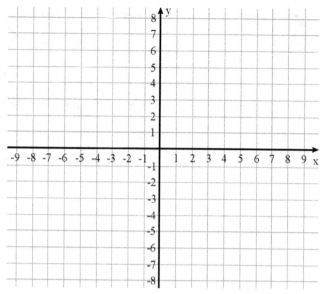

7. The points $(-7, -6)$, $(-5, -2)$, and $(-1, -4)$ are vertices of a triangle.

a. Draw the triangle.

b. Reflect it in the x-axis. Then move the already reflected triangle 5 units to the right.

c. Draw the new triangle and write the coordinates of the new vertices.

(_____ , _____)

(_____ , _____)

(_____ , _____)

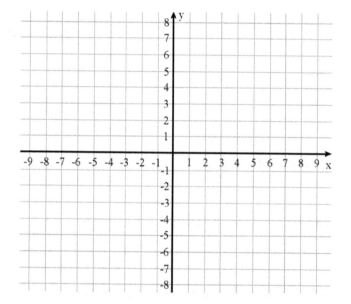

Grade 6, Chapter 9

End-of-Chapter Test

Instructions to the student:

A calculator is not allowed. Answer each question in the space provided.

Instructions to the teacher:

My suggestion for grading is as follows. The total is 26 points. You can give partial points for partial solutions.

Divide the student's score by 26 to get a percent score. For example, if the student scores 21, divide $21 \div 26$ with a calculator, getting 0.80769230... The percent score is 81%.

Question #	Max. points	Student score
1	4 points	
2	3 points	
3	4 points	
4	3 points	
5	2 points	

Question #	Max. points	Student score
6	2 points	
7	4 points	
8	2 points	
9	2 points	
TOTAL	26 points	/ 26

Test - Chapter 9

*A calculator **is not** allowed.*

1. Measure what you need from the shape, and find
   its area...

   a. ...in square centimeters, to the nearest
   square centimeter.

   b. ...in square millimeters, to the nearest
   hundred square millimeters.

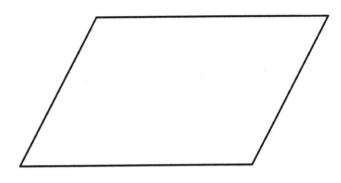

2. Find the area of the quadrilateral
   in square units.

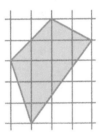

3. **a.** What is this shape called?

   b. Find its area.

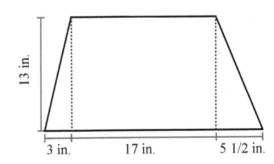

4. The dimensions of this box are 2 ft ï¿½ 1.5 ft ï¿½ 1.5 ft.
   What is the total area of the bottom and side faces of the box
   (ie. not including the top)?

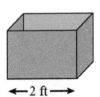

5. The edges of each little cube measure 1/4 in.
   What is the total volume of the figure?

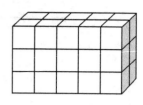

6. A book about how to raise ducks measures 6 1/2 in. × 8 in. ï¿½ 3/8 in.
   What is the volume of one book?

7. **a.** What solid can be built from this net?

   b. Calculate its surface area, if each side of the bottom square
   measures 5 in. and the height of each triangle is 4 1/8 in.

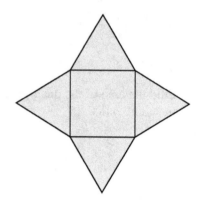

8. What is this solid called?

   Sketch its net.

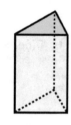

9. The vertices of a triangle are
   (1, 0), (−2, −4) and (−3, −3).
   Find the area of the triangle.

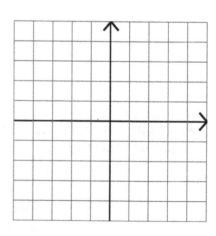

Grade 6, Chapter 10

End-of-Chapter Test

Instructions to the student:

A calculator is not allowed. Answer each question in the space provided.

Instructions to the teacher:

My suggestion for grading is as follows. The total is 25 points. You can give partial points for partial solutions.

Multiply the student's score by 4 to get a percent score. For example, if the student scores 21, multiply $4 \times 22 = 88$. The percent score is 88%.

Question #	Max. points	Student score
1	8 points	
2a	2 points	
2b	2 points	
2c	2 points	
2d	1 point	

Question #	Max. points	Student score
3a	3 points	
3b	1 point	
3c	1 point	
4	5 points	
TOTAL	25 points	/ 25

Test - Chapter 10

*A calculator **is** allowed.*

1. Calculate the mean, median, mode, and range—if possible—for these data sets.

    a. 12, 15, 11, 18, 20, 15, 16

    mean _____ median _____ mode _____ range _____

    b. duck, cow, horse, horse, horse, cat, cat, dog, dog, dog

    mean _____ median _____ mode _____ range _____

2. The following are the points for two math quizzes for a 7th grade class.

    a. Make bar graphs from the data.

    b. Describe the shape of each distribution.

    c. Choose a measure of center to describe the distributions, and determine its value for both quizzes.

    d. Which quiz went better overall?

Quiz 1	
Points	**Students**
5	7
6	8
7	6
8	3
9	0
10	0

Quiz 2	
Points	**Students**
5	1
6	2
7	5
8	8
9	5
10	3

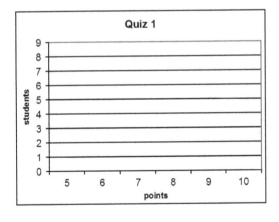

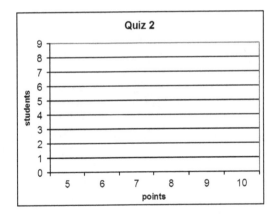

Quiz 1:

Shape of the distribution: _____

_____

Measure of center: _____

Quiz 2:

Shape of the distribution: _____

_____

Measure of center: _____

3. **a.** Make a stem-and-leaf plot of this data.

114 128 132 127 122 127 130 119 120 121 125
(Results of a high jump contest boys, in centimeters)

Stem	Leaf

b. Find the median.

c. What is the interquartile range?

4. Make a boxplot from this data:

89 92 95 96 99 103 105 106 106 109 109 110 112 114 117 118 124

(birth weight in grams of Momma Cat's three litters of kittens)

Grade 6 End-of-the-Year Test

Instructions

This test is quite long, because it contains lots of questions on all of the major topics covered in the *Math Mammoth Grade 6 Complete Curriculum*. Its main purpose is to be a diagnostic test—to find out what the student knows and does not know. The questions are quite basic and do not involve especially difficult word problems.

Since the test is so long, I do not recommend that you have the student do it in one sitting. You can break it into 3-5 parts and administer them on consecutive days, or perhaps on morning/evening/morning/evening. Use your judgment.

A calculator is not allowed, except on the page about measuring units.

The test is evaluating the student's ability in the following content areas:

- exponents, expanded form, and rounding
- writing and simplifying expressions
- the distributive property
- the concept of an equation and solving simple equations
- the concept of inequality
- all operations with decimals
- conversions between measuring units
- basic ratio concepts
- the concept of percentage, finding percentages, finding the percent of number
- prime factorization, the greatest common factor, and the least common multiple
- division of fractions
- basic concepts related to integers
- addition and subtraction of integers
- the area of triangles, parallelograms, and polygons
- surface area and nets
- the volume of rectangular prisms
- describing statistical distributions
- measures of center
- statistical graphs

Instructions to the student:
Do not use a calculator. Answer each question in the space provided.

Instructions to the teacher:
In order to continue with the *Math Mammoth Grade 7 Complete Worktext,* I recommend that the student score a minimum of 80% on this test, and that the teacher or parent review with the student any content areas in which the student may be weak. Students scoring between 70% and 80% may also continue with grade 7, depending on the types of errors (careless errors or not remembering something, versus a lack of understanding). Use your judgment.

My suggestion for points per item is as follows. The total is 194 points. A score of 155 points is 80%.

Question #	Max. points	Student score
Basic Operations		
1	2 points	
2	3 points	
3	2 points	
4	2 points	
	subtotal	/ 9
Expressions and Equations		
5	4 points	
6	2 points	
7	2 points	
8	1 point	
9	2 points	
10	2 points	
11	2 points	
12	2 points	
13	2 points	
14	2 points	
15	1 point	
16	2 points	
17	2 points	
18	2 points	
19	4 points	
	subtotal	/ 32
Decimals		
20	2 points	
21	2 points	
22	1 point	
23	2 points	
24	2 points	
25	1 point	
26	2 points	
27	2 points	
28a	1 point	
28b	2 points	
29	3 points	
	subtotal	/ 20

Question #	Max. points	Student score
Measuring Units		
30	3 points	
31	1 point	
32	2 points	
33	3 points	
34	6 points	
35	4 points	
	subtotal	/ 19
Ratio		
36	2 points	
37	2 points	
38	2 points	
39	2 points	
40	2 points	
41	2 points	
42	2 points	
	subtotal	/ 14
Percent		
43	3 points	
44	4 points	
45	2 points	
46	2 points	
47	2 points	
	subtotal	/13

Question #	Max. points	Student score
Prime Factorization, GCF, and LCM		
48	3 points	
49	2 points	
50	2 points	
51	2 points	
52	2 points	
	subtotal	/11
Fractions		
53	3 points	
54	2 points	
55	2 points	
56	2 points	
57	3 points	
58	3 points	
	subtotal	/15
Integers		
59	2 points	
60	2 points	
61	2 points	
62	4 points	
63	5 points	
64	6 points	
65	4 points	
	subtotal	/25

Question #	Max. points	Student score
Geometry		
66	1 point	
67	1 point	
68	3 points	
69	4 points	
70	2 points	
71a	1 point	
71b	3 points	
72	4 points	
73a	2 points	
73b	2 points	
	subtotal	/23
Statistics		
74a	2 points	
74b	1 point	
74c	2 points	
75a	1 point	
75b	1 point	
76a	2 points	
76b	1 point	
76c	1 point	
76d	2 points	
	subtotal	/13
	TOTAL	/194

Math Mammoth End-of-the-Year Test - Grade 6

Basic Operations

1. Two kilograms of ground cinnamon is packaged into bags containing 38 g each. There will also be some cinnamon left over. How many bags will there be?

2. Write the expressions using an exponent. Then solve.

   a. $2 \times 2 \times 2 \times 2 \times 2$

   b. five cubed

   c. ten to the seventh power

3. Write in normal form (as a number).

   a. $7 \times 10^7 + 2 \times 10^5 + 9 \times 10^0$

   b. $3 \times 10^8 + 4 \times 10^6 + 5 \times 10^5 + 1 \times 10^2$

4. Round to the place of the underlined digit.

   a. $6,29\underline{9},504 \approx$ _____

   b. $6,609,\underline{9}42 \approx$ _____

Expressions and Equations

5. Write an expression.

   a. 2 less than s

   b. the quantity $7 + x$, squared

   c. 5 times the quantity $y - 2$

   d. the quotient of 4 and x^2

6. Evaluate the expressions when the value of the variable is given.

a. $40 - 8x$ when $x = 2$	**b.** $\dfrac{65}{p} \cdot 3$ when $p = 5$

7. Write an expression for each situation.

    a. You bought m yogurt cups at \$2 each and paid with \$50.
       What is your change?

    b. What is the area of a square with side length s?

8. Write an expression for the total length of the line segments, and simplify it.

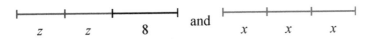

    z z 8   and x x x

9. Write an expression for the perimeter of the figure, and simplify it.

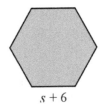

$s + 6$

10. Write an expression for the area of the figure, and simplify it.

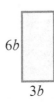

$6b$

$3b$

11. Simplify the expressions.

a. $9x - 6x$	**b.** $w \cdot w \cdot 7 \cdot w \cdot 2$

12. Multiply using the distributive property.

a. $7(x + 5) =$	**b.** $2(6p + 5) =$

13. Find the missing number in the equations.

a. ____ $(6x + 5) = 12x + 10$	**b.** $5(2h +$ ____ $) = 10h + 30$

14. Solve the equations.

a. $\dfrac{x}{31} = 6$	**b.** $a - 8.1 = 2.8$

15. Which of the numbers 0, 1, 2, 3 or 4 make the equation $\dfrac{8}{y^2} = 2$ true?

16. Write an equation EVEN IF you could easily solve the problem without an equation! Then solve the equation. The value of a specific number of quarters is 1675 cents. How many quarters are there?

17. Write an inequality for each phrase. You will need to choose a variable to represent the quantity in question.

   a. Eat at most 5 pieces of bread.

   b. You have to be at least 21 years of age.

18. Write an inequality that corresponds to the number line plot.

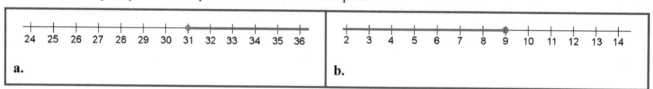

a.

b.

19. A car is traveling with a constant speed of 80 kilometers per hour. Consider the variables of time (t), measured in hours, and the distance traveled (d), measured in kilometers.

   a. Fill in the table.

t (hours)	0	1	2	3	4	5	6
d (km)							

   b. Plot the points on the coordinate grid.

   c. Write an equation that relates t and d.

   d. Which of the two variables is the independent variable?

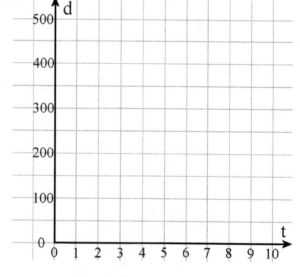

Decimals

20. Write as decimals.

   a. 13 millionths   **b.** 2 and 928 ten-thousandths

21. Write as fractions or mixed numbers.

   a. 0.00078   **b.** 2.000302

22. Find the value of the expression $x + 0.07$ when x has the value 0.0002.

23. Calculate mentally.

a. $0.8 \div 0.1 =$	**b.** $0.06 \times 0.008 =$

24. **a.** Estimate the answer to 7.1×0.0058.

   b. Calculate the exact answer.

25. What number is 22 ten-thousandths more than 1 1/2?

26. Multiply or divide.

a. $10^5 \times 0.905 =$	**b.** $24 \div 10^4 =$

27. Divide, and give your answer as a decimal. If necessary, round the answers to three decimal digits.

a. $175 \div 0.3$	**b.** $\dfrac{2}{9}$

28. Annie bought 3/4 kg of cocoa powder, which cost $12.92 per kg.

   a. Estimate the cost.

   b. Find the exact amount she had to pay.

29. Alyssa and Anna bought three toy cars for their three cousins from a store on line. The price for one car was $3.85. A shipping fee of $4.56 was added to the total cost. The two girls shared the total cost equally. How much did each girl pay?

Measuring Units   *A calculator is allowed in this section.*

1 mile = 5,280 feet	1 ton = 2,000 lb	1 gal = 4 qt
1 mile = 1,760 yards	1 lb = 16 oz	1 qt = 2 pt
		1 pt = 16 fl. oz

30. Convert to the given unit. Round your answers to two decimals, if needed.

a. 178 fl. oz. = _____ qt	**b.** 0.412 mi = _____ ft	**c.** 1.267 lb = _____ oz

31. How many miles is 60,000 inches?

32. A big coffee pot makes 2 quarts of coffee.
    How many 6-ounce servings can you get from that?

33. A pack of 36 milk chocolate candy bars costs $23.20. Each bar weighs 1.6 oz.
    Calculate how much one pound of these chocolate bars would cost (price per pound).

34. Convert the measurements. You can write the numbers in the place value charts to help you.

a. 39 dl = _____ L	**b.** 15,400 mm = _____ m
c. 7.5 hm = _____ cm	**d.** 597 hl = _____ L
e. 7.5 hg = _____ kg	**f.** 32 g = _____ cg

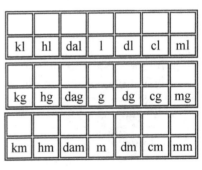

kl	hl	dal	l	dl	cl	ml

kg	hg	dag	g	dg	cg	mg

km	hm	dam	m	dm	cm	mm

35. **a.** One brick is 215 mm long. How many of these bricks,
    put end to end, will cover a 5.15 meter wall?

    b. Calculate the answer to the previous question again,
    assuming 1 cm of mortar is laid between the bricks.

Ratio

36. **a.** Draw a picture where there are a total of ten squares,
   and for each two squares, there are three triangles.

   b. Write the ratio of squares to all triangles,
   and simplify this ratio to the lowest terms.

37. Write ratios of the given quantities. Then, simplify the ratios. You will need to *convert* one quantity
   so it has the same measuring unit as the other.

a. 3 kg and 800 g	**b.** 2.4 m and 100 cm

38. Express these rates in the lowest terms.

a. $56 : 16 kg	**b.** There are six teachers for every 108 students.

39. Change to unit rates.

a. $20 for five T-shirts	**b.** 45 miles in half an hour

40. **a.** It took 7 hours to mow four equal-size lawns. At that rate, how many lawns could be mowed
   in 35 hours? You can use the table below to help.

Lawns					
Hours					

   b. What is the unit rate?

41. Joe and Mick also worked on a project unequally. They decided to divide
   their pay in a ratio of 3:4 (3 parts for Joe, 4 parts for Mick). The total pay was $180.
   Calculate how much Mick got.

42. Use the given ratios to convert the measuring units. If necessary, round the answers to three decimal digits.

a. Use $1 = \dfrac{1.6093 \text{ km}}{1 \text{ mi}}$ and convert 7.08 miles to kilometers. 7.08 mi =	
b. Use $1 = \dfrac{1 \text{ qt}}{0.946 \text{ L}}$ and convert 4 liters to quarts. 4 L =	

Percent

43. Write as percentages, fractions, and decimals.

a. _____ % $= \dfrac{35}{100} =$ _____	**b.** 9% $= \dfrac{}{} =$ _____	**c.** _____ % $= \dfrac{}{} = 1.05$

44. Fill in the table, using mental math.

	510
1% of the number	
5% of the number	
10% of the number	
30% of the number	

45. A pair of roller skates is discounted by 40%. The normal price is $65.
    What is the discounted price?

46. A store has sold 90 notebooks, which is 20% of all the notebooks they had.
    How many notebooks did the store have at first?

47. Janet has read 17 of the 20 books she borrowed from the library.
    What percentage of the books she borrowed has she read?

Prime Factorization, GCF, and LCM

48. Find the prime factorization of the following numbers.

a. 45 / \	b. 78 / \	c. 97 / \

49. Find the least common multiple of these pairs of numbers.

a. 2 and 8	b. 9 and 6

50. Find the greatest common factor of the given number pairs.

a. 30 and 16	b. 45 and 15

51. List three different multiples of 28 that are more than 100 and less than 200.

52. First, find the GCF of the numbers. Then factor the expressions using the GCF.

a. The GCF of 18 and 21 is _____ 18 + 21 = ____ · ____ + ____ · ____ = ____ (____ + ____)
b. The GCF of 56 and 35 is _____ 56 + 35 = ____ (____ + ____)

Fractions

53. Solve.

a. $\dfrac{4}{5} \div \dfrac{1}{5}$	**b.** $3\dfrac{1}{8} \div 1\dfrac{1}{2}$	**c.** $4 \div \dfrac{5}{7}$

54. Write a division sentence, and solve.

How many times does go into  ?

55. Write a real-life situation to match this fraction division: $1\dfrac{3}{4} \div 3 = \dfrac{7}{12}$

56. How many 3/4-cup servings can you get from 7 1/2 cups of coffee?

57. A rectangular room measures 12 1/2 feet by 15 1/3 feet. It is divided into three equal parts. Calculate the area of one of those parts.

58. The perimeter of a rectangular screen is 15 1/2 inches, and the ratio of its width to its height is 3:2. Find the width and height of the screen.

Integers

59. Compare the numbers, writing < or > in the box.   **a.** 0 ☐ −3   **b.** −2 ☐ −8

60. Write a comparison to match each situation (with < or >).

    a. The temperature −7°C is warmer than −12°C.

    b. Harry has $5. Emily owes $5.

61. Find the difference between the two temperatures.

    a. −13°C and 10°C   **b.** −9°C and −21°C

62. Write using mathematical symbols, and simplify (solve) if possible.

    a. The opposite of 7   **b.** the absolute value of −6

    c. the absolute value of 5   **d.** the absolute value of the opposite of 6

63. **a.** Plot the point (−5, 3).

    b. Reflect the point in the x-axis .

    c. Now, reflect the point you got in (b) in the y-axis.

    d. Join the three points with line segments.
       What is the area of the resulting triangle?

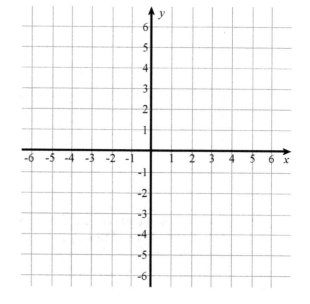

64. Draw a number line jump for each addition or subtraction sentence, and solve.

    a. −2 + 5 = _____

    b. −2 − 4 = _____

    c. −1 − 5 = _____

65. Write an addition or subtraction in the box to match each situation, and fill in the blanks.

    a. Elijah has saved $10. He wants to buy shoes for $14.
       That would make his money situation to be _____.

    b. A fish was swimming at the depth of 2 m. Then it sank 1 m.
       Now he is at the depth of _____ m.

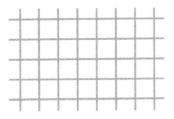

Geometry

66. Draw in the grid a right triangle with a base of 4 units and a height of 3 units.

    Calculate its area.

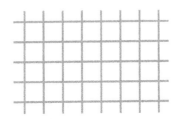

67. Draw in the grid a parallelogram with an area of 15 square units.

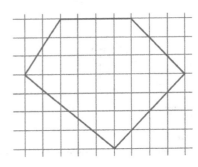

68. Find the area of this polygon, in square units.

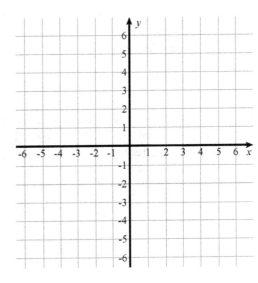

69. Draw a quadrilateral in the grid with vertices
    $(-5, 5)$, $(-5, -3)$, $(2, -1)$, and $(2, 4)$.

    What is the quadrilateral called?

    Find its area.

70. Name this solid. Draw a sketch of its net.

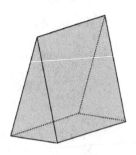

71. **a.** Name the solid that can be built from this net.

b. Calculate its surface area.

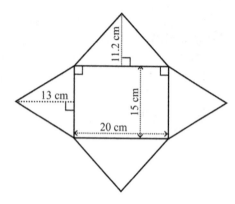

72. The edges of each little cube measure **1/2 cm**. What is the total volume of these figures, in cubic units?

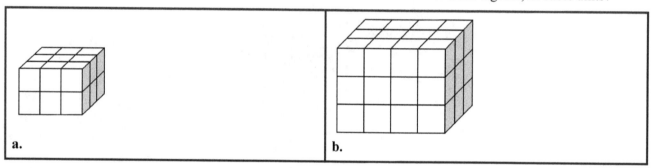

a.

b.

73. A box containing a construction toy measures 1 3/4 in. by 8 1/2 in. by 6 inches.

   a. Calculate its volume.

   b. Would 12 of these boxes fit into a crate with the inside
      measurements of 1 ft by 1 ft by 1 ft?
      Justify your reasoning.

Statistics

74. **a.** Make a stem-and-leaf plot of this data.

55 59 61 62 64 65 65 68 69 70 72 74 77 83 89 94

(The ages of people in a senior chess club)

b. Find the median.

c. Find the interquartile range.

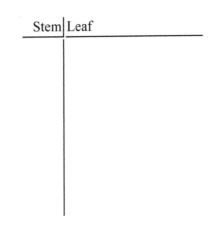

75. **a.** Describe the shape of this distribution.

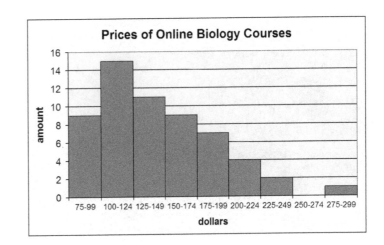

b. Which measure of center would be best to describe this distribution?

76. **a.** Create a dot plot from this data.

9 10 5 6 4 8 7 3 8 1 7 7 5 7 8 9 5 6 6 7

(points on a math quiz of a group of students)

b. Describe the shape of the distribution.

c. Describe the spread of the data.

d. Choose a measure of center to describe the data, and determine its value.

Using the Cumulative Reviews

The cumulative reviews practice topics in various chapters of the Math Mammoth complete curriculum, up to the chapter named in the review. For example, a cumulative review for chapters 1-6 may include problems matching chapters 1, 2, 3, 4, 5, and 6. The cumulative review lesson for chapters 1-6 can be used any time after the student has studied the curriculum through chapter 6.

These lessons provide additional practice and review. The teacher should decide when and if they are used. The student doesn't have to complete all the cumulative reviews. I recommend using at least three of these reviews during the school year. The teacher can also use the reviews as diagnostic tests to find out what topics the student has trouble with.

Math Mammoth complete curriculum also includes an easy worksheet maker, which is the perfect tool to make more problems for children who need more practice. The worksheet maker covers most topics in the curriculum, excluding word problems. Most people find it to be a very helpful addition to the curriculum.

The download version of the curriculum comes with the worksheet maker, and you can also access the worksheet maker online at

https://www.mathmammoth.com/private/Make_extra_worksheets_grade6.htm

Cumulative Review, Grade 6, Chapters 1-2

1. **a.** A boat is traveling at the constant speed of 24 kilometers per hour. Fill in the table.

Distance		12 km		24 km		216 km
Time	10 min		50 min		5 1/2 hours	

    b. How long will it take for the boat to travel 360 kilometers?

2. **a.** Estimate the answer to $234 \times 1{,}091$.

    b. Multiply $234 \times 1{,}091$ (in the space on the right, or in your notebook).

    c. Now, estimate the answer to 2.34×1.091.

    d. Based on your answers to (b) and (c), what is 2.34×1.091?

3. **a.** Write a subtraction equation where the minuend is 56, the difference is 17, and the subtrahend is the unknown y. Then solve for y.

    b. Write a division equation where the quotient is 60, the divisor is 15, and the dividend is unknown. Solve it.

4. Divide mentally in parts.

    a. $\dfrac{636}{6}$ **b.** $\dfrac{824}{4}$ **c.** $\dfrac{5{,}607}{7}$ **d.** $\dfrac{1{,}224}{12}$

5. Find the value of these expressions.

a. $100 - 100 \div 4 \cdot 2$	**b.** $3^3 \div (4 + 5)$
c. $(2 + 6)^2 - (25 - 5)$	**d.** $\dfrac{12^2 + 9}{5 \cdot 3}$

6. Evaluate the expressions when the value of the variable is given.

a. $3x - 12$ when $x = 5$	b. $\dfrac{y}{3} + 4$ when $y = 24$

7. A rectangle's width is w and its length is l. Which expression tells us the perimeter of the rectangle?

   a. lw **b.** $2l + 2w$ **c.** $\dfrac{l}{w}$ **d.** $l + w + l$ **e.** $l + w$

8. Write an expression for the area (A) or volume (V) using an *exponent*, and solve.

a. A square with sides 11 cm in length: A = _____	b. A cube with edges that are all 4 ft long: V = _____

9. The perimeter of a square is 64 cm. What is its area?

10. One pair of shoes costs \$48.60, and another pair costs 2/3 of that price.
   Alyssa bought both pairs. Find her change from \$100.

11. Divide. If the division is not exact, give your answer to three decimals.

a. $17\overline{)267087}$	b. $15\overline{)8}$	c. $3\overline{)0.13}$

Cumulative Review, Grade 6, Chapters 1-3

1. Four parents shared the cost of $207.48 for hosting a parent meeting
   in this way: one parent paid half of the cost, and the rest was divided
   equally between the rest of the parents. How much did each parent pay?
   Hint: you can draw a bar model to help.

2. Write in normal form (as a number).

   a. $3 \cdot 10^8 + 2 \cdot 10^7 + 9 \cdot 10^6 + 3 \cdot 10^2$

   b. $1 \cdot 10^6 + 5 \cdot 10^4 + 3 \cdot 10^0$

3. Round to the place of the underlined digit. Be careful with the nines!

   a. $5,69\underline{9},528 \approx$ _____

   b. $219,99\underline{7},101 \approx$ _____

   c. $8\underline{2},788,000 \approx$ _____

   d. $3,999,9\underline{9}2,567 \approx$ _____

4. Evaluate the expression for the given values of the variable x.

Variable	Expression $\dfrac{x^2}{3}$	Value
$x = 1$	$\dfrac{1^2}{3}$	$\dfrac{1}{3}$
$x = 2$		

Variable	Expression $\dfrac{x^2}{3}$	Value
$x = 3$		
$x = 5$		

5. Write an expression for each scenario, and then find its value.

   a. The sum of 12 and 56 divided by 4.

   b. The quotient of 8 and the quantity 4 to the third power.

6. Simplify the expressions.

a. $c \cdot c \cdot c \cdot 8 \cdot c$	**b.** $7c - 2c + 8$
c. $t + t + t + 3 - 2t$	**d.** $2x^2 + 5 + 11x^2 + 8$

7. Write an expression for each situation.

 a. Anna has m marbles. She gave 1/3 of them to her friend.
  How many marbles did her friend get?

 b. Sadie is s years old. Fanny is 6 years younger than Sadie.
  How old is Fanny?

 c. How old will Sadie be in 5 years?

 d. How old will Fanny be in 5 years?

8. Solve these equations.

a. $\quad 7x + 2x \;=\; 54$	**b.** $\quad 8r - 3r \;=\; 40$	**c.** $\quad t \div 50 \;=\; 5 + 6$
$\quad\quad\quad\quad =$	$\quad\quad\quad\quad =$	$\quad\quad\quad\quad =$
$\quad\quad\quad\quad =$	$\quad\quad\quad\quad =$	$\quad\quad\quad\quad =$
$\quad\quad\quad\quad =$	$\quad\quad\quad\quad =$	$\quad\quad\quad\quad =$
d. $\quad w - 88 \;=\; 20 \cdot 60$	**e.** $\quad 2x - 6 \;=\; 16$	**f.** $\quad 8x + 17 \;=\; 81$
$\quad\quad\quad\quad =$	$\quad\quad\quad\quad =$	$\quad\quad\quad\quad =$
$\quad\quad\quad\quad =$	$\quad\quad\quad\quad =$	$\quad\quad\quad\quad =$
$\quad\quad\quad\quad =$	$\quad\quad\quad\quad =$	$\quad\quad\quad\quad =$

9. Think of the distributive property "backwards," and factor these sums.

a. $16y + 12 =$ ____ (____ $y +$ ____)	**b.** $9x + 9 =$ ____ (____ $+$ ____)
c. $54c + 24 =$ ____ (____ $+$ ____)	**d.** $15a + 45 =$ ____ (____ $+$ ____)

10. Solve the equations.

a. $x + 4.5039 = 7$	**b.** $0.938208 - x = 0.047$	**c.** $2x = 6.0184$

Cumulative Review, Grade 6, Chapters 1-4

1. Write the statements as equations. Then solve the equations.

a. The quotient of a secret number and 11 is equal to 12.	**b.** The sum of 3, 5, and a certain number is 105.

2. Solve the equations.

a. $x \div 6 = 40 + 50$	**b.** $1{,}000 - x = 40 \cdot 6$	**c.** $8x + 2x = 15 \cdot 6$

3. The numbers below are prices for sets of 12 colored pencils from seven different stores.

$3.89 $3.99 $4.45 $3.79 $4.10 $4.19 $4.02

a. Find the average price.

b. How much will the teacher save if she buys 100 sets of the pencils at the cheapest price as compared to the most expensive price?

4. Calculate.

a.	**b.**	**c.**
$10 \cdot 0.009 =$	$40 \cdot 0.08 =$	$0.1 \cdot 0.2 \cdot 0.3 =$
$0.5 \cdot 0.6 =$	$1{,}000 \cdot 1.2 =$	$0.11 \cdot 0.02 =$
d.	**e.**	**f.**
$10 \div 0.2 =$	$0.075 \div 0.025 =$	$2.36 \div 2 =$
$0.6 \div 0.2 =$	$0.3 \div 0.02 =$	$0.0045 \div 5 =$

5. Write the amounts in basic units (meters, grams, or liters).

a. 6 kg = _____ g	**b.** 7 dam = _____ m	**c.** 7 kl = _____ L
5 dl = _____ L	5 hl = _____ L	50 mg = _____ g
5 mm = _____ m	30 cg = _____ g	8 cm = _____ m

6. We often compare the size of people and animals by comparing their weights.
   Tim weighs 45 kg, and a grasshopper weighs 3,000 mg.

   a. How many times more does Tim weigh than the grasshopper?

   b. Assuming that they were somehow packaged to carry,
   could you carry the weight of a thousand grasshoppers?

7. Elaine paid 1/5 of her salary for taxes. Then she paid 1/6 of what was left for rent.
   Then she had $1,000 left. How much was her salary?

					rent

				tax

8. Divide. If necessary, round your answer to three decimal digits.

a. $45.7 \div 0.02$	**b.** $928 \div 0.003$	**c.** $\dfrac{5}{8}$

Cumulative Review, Grade 6, Chapters 1-5

A calculator is allowed only in the last problem.

1. Write as decimals.

   a. 392 hundred-thousandths

   b. 5 and 15 ten-thousandths

   c. 23 millionths

   d. 12 and 12 thousandths

2. Write as fractions.

   a. 0.000016

   b. 2.9381

   c. 0.39402

3. Find the value of the expression $y - 0.05$ when

a. $y = 1$	**b.** $y = 0.1$	**c.** $y = 1.1$

4. Round to...

	2.97167	**0.046394**	**2.33999**	**1.199593**
the nearest tenth				
the nearest thousandth				

5. Multiply both the dividend and the divisor by the same number, so that the divisor will be a whole number. Then divide mentally.

a. $\dfrac{5.6}{0.4} = \underline{\quad} = $	**b.** $\dfrac{4}{0.02} = \underline{\quad} = $	**c.** $\dfrac{0.9}{0.003} = \underline{\quad} = $

6. When 1,200 people were polled about their favorite foods, 320 said they liked mashed potatoes best.

   a. Write a ratio, and simplify it to the lowest terms.

   b. Assuming the same ratio holds true in another group of 150 people, how many of those people can we expect to have mashed potatoes as their favorite food?

7. Fill in the missing numbers to form equivalent rates.

a. $\dfrac{14\ km}{20\ min} = \dfrac{}{5\ min} = \dfrac{}{45\ min}$

b. $\dfrac{}{8\ bottles} = \dfrac{}{1\ bottle} = \dfrac{\$42}{10\ bottles}$

8. You need 2 kg of fertilizer for every 120 m² of lawn.
   How much fertilizer would you need for a rectangular 15 m by 20 m lawn?

9. Calculate the values of y according to the equation
   $y = 2x - 4$.

x	2	3	4	5	6	7
y						

Then plot the points.

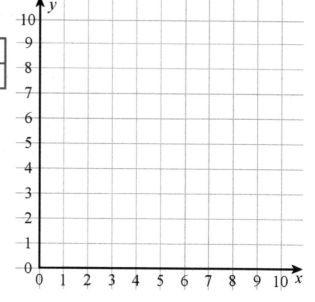

10. Two-thirds of a stick is 50 cm long.
    How long is the whole stick?

11. Marsha has 2 gallons of punch, which she is pouring into 6-oz servings.
    How many servings will she be able to get?

12. The children Hannah, 120 cm, and Erica, 1.05 m, stand on stools to see how tall they are.
    At what height is the top of their head, if the children stand on stools with the heights of:

    a. 3.1 dm

    b. 550 mm

    c. 45 cm

13. Convert the given distances into metric units. Round the numbers to one
    decimal place. *Use a calculator* and the conversion factors at the right. →

    Every afternoon Erica bicycles 5 miles (_____ km) to the horse ranch.

    Erica takes care of a horse that is 15 *hands*, or 60 inches (_____ m), tall.

    She likes to go riding on a trail that is 4 mi 500 ft (_____ km) long.

    > 1 inch = 2.54 cm
    > 1 foot = 0.3048 m
    > 1 mile = 1.6093 km

Cumulative Review, Grade 6, Chapters 1-6

A calculator is allowed only in the last problem.

1. Write as percentages. If necessary, round your answers to the nearest percent.

    a. 4/5

    b. 17/20

    c. 5/11

2. Write the fractions from the previous problem as decimals.

    a. 4/5 =   **b.** 17/20 =   **c.** 5/11 =

3. A store got a shipment of 155 calculators.
    The ratio of basic calculators to scientific calculators was 4:1.

    a. Draw a model to represent the situation.

    b. What fractional part of the calculators were basic calculators?

    c. What percentage of the calculators were basic calculators?

    d. How many scientific calculators were there?

4. Mike rides his bike at a constant speed of 20 km/h. Fill in the table.

Distance				16 km	20 km	24 km			
Time	6 min	12 min	15 min		1 hour		2 hours	3 hours	3 1/2 hours

5. One flash drive costs $25 and another costs 15% more.
    Find the total cost of buying both.

6. Think of the distributive property "backwards," and factor these sums.

a. $32t + 8 =$ ____ (____ + ____)	**b.** $8 + 12x =$ ____ (____ + ____)
c. $15y + 45 =$ ____ (____ + ____)	**d.** $35 + 42w =$ ____ (____ + ____)

7. Grace got 35 points out of 40 in a test. Convert her test score into a percentage.

8. Jack gave 4/5 of his 90 toy cars to his cousins.
   Then he divided the rest of his cars equally with his brother.
   How many cars does Jack have now?

9. Multiply or divide the decimals by the powers of ten.

a. $10^4 \times 0.092 =$	**b.** $1,000 \times 0.0004 =$
c. $456.29 \div 1,000 =$	**d.** $63 \div 10^5 =$

10. Change into the basic unit (meter, liter, or gram).

   a. 1534 cm   **b.** 334 ml   **c.** 0.9 kg

11. Write an equation for each situation EVEN IF you could easily solve the problem without an equation!
    Then solve the equation.

   a. A camera costs $85 more than a camera bag.
      If the camera costs $162, how much does the camera bag cost?

   b. Jennifer purchased a set of 8 towels for $52.
      How much did one towel cost?

12. If three shirts cost $14.10, then how much do seven shirts cost?

13. Convert into the given units. Round your answers to 2 decimals if needed.

a. 79 oz = _____ lb _____ oz	**c.** 7.82 qt = _____ gal	**e.** 2.54 lb = _____ oz
b. 4 ft 11 in = _____ in	**d.** 0.265 mi = _____ yd	**f.** 6.8 ft = _____ ft _____ in

Cumulative Review, Grade 6, Chapters 1-7

A calculator is not allowed.

1. Find the least common multiple of these pairs of numbers.

a. 2 and 8	**b.** 6 and 9

2. Find the greatest common factor of the given number pairs.

a. 14 and 15	**b.** 48 and 60

3. Draw two rectangles, side by side, to represent the sum $45 + 27$.

4. A container of ice cream contains 2 quarts of ice cream.
   This is divided equally between 9 people.
   How much will each person get (in ounces)?

5. Fill in.

   a. 11^2 gives us the _____ of a _____ with a side length of _____ units.

   b. 3×5^2 gives us the _____ of _____ _____ with a side length of _____ units.

   c. 4×0.4^3 gives us the _____ of _____ _____ with an edge length of _____ units.

6. Write in normal form (as a number).

a. $2 \times 10^7 + 6 \times 10^6 + 2 \times 10^4$	**b.** $1 \times 10^9 + 2 \times 10^5 + 8 \times 10^2 + 7 \times 10^0$

7. Factor the following numbers into their prime factors.

a. 99 /\	b. 112 /\	c. 200 /\

8. Chris has two kinds of containers for gasoline. The larger ones
   hold 8.5 liters, and the smaller ones hold 60% of that amount.
   What is the total capacity of three large and four smaller containers?

9. Samantha and George got paid $100 for working together on a project.
   Since Samantha had worked 5 hours and George only 3 hours, they
   decided it would be fair to divide the pay in a ratio of 5:3.
   How much more did Samantha earn than George?

10. Write an expression.

   a. the quotient of 6 and $7s$

   b. Subtract $2x$ from 11

   c. the sum of x and 2, squared

   d. the quantity $5m$, cubed

   e. $2t^2$ divided by the difference of s and 1

   f. y less than 18

11. Write an equation to match the bar model, and solve it.

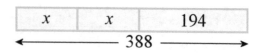

12. Multiply.

a. $4 \times 0.7 =$ _____	c. $3 \times 1.06 =$ _____	e. $10^5 \times 0.08 =$ _____
b. $50 \times 0.003 =$ _____	d. $100 \times 0.009 =$ _____	f. $40 \times 0.004 =$ _____

Cumulative Review, Grade 6, Chapters 1-8

A calculator is not allowed.

1. Find the reciprocal.

a. $1\frac{1}{23}$	b. $3\frac{2}{11}$	c. 79	d. 100	e. $\frac{3}{1000}$

2. Divide.

a. $\frac{6}{7} \div \frac{1}{7}$	b. $1\frac{9}{20} \div \frac{3}{20}$	c. $5 \div \frac{1}{3}$	d. $7 \div 1\frac{2}{5}$

3. A sheet of stickers has 48 stickers that each measure 1 1/4 in. by 1 1/4 in.
Little Hannah starts to stick them on the front cover of a notebook,
side by side. The notebook measures 5 1/2 in. by 8 1/2 in.
How many stickers can she fit on the cover?

4. Fill in the table.

Expression	the terms in it	coefficient(s)	Constants
$2x + 3y$			
$0.9s$			
$2a^4c^5 + 6$			
$\frac{1}{6}f$			

5. The table lists the quantities of some of the ingredients for different-sized batches of a certain cake recipe. Fill in the table.

Serves (people)	6	12	18	24	30
butter		1/2 cup			
sugar		1 cup			
eggs		2			
flour		1 1/2 cups			

6. If you make enough cakes for 100 people, how much butter, sugar, eggs, and flour are needed? (You can use the table above.)

7. Find the value of the expressions.

a. $900 - \dfrac{1}{6} \cdot 72$	**b.** $23 + 3^4$	**c.** $\dfrac{100^3}{100^2}$

8. Marie's age is 4/7 of her brother Tom's age. Tom is 9 years older than Marie. (You can draw a bar diagram to help.)

   a. How old is Marie? Tom?

   b. What is the ratio of Marie's age to Tom's age?

9. Write an expression for both the area *and* perimeter of each shape, in simplified form.

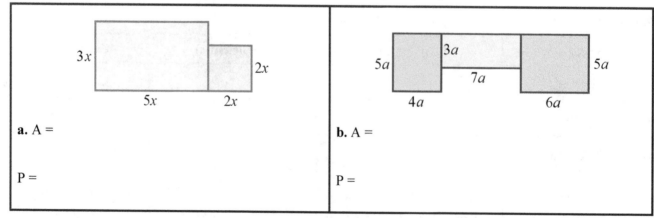

a. A =

P =

b. A =

P =

Cumulative Review, Grade 6, Chapters 1-9

A calculator is not allowed.

1. Move these points four units to the *left*:

   $(-5, 1) \rightarrow ($ ___ , ___ $)$

   $(-2, -3) \rightarrow ($ ___ , ___ $)$

   $(3, -7) \rightarrow ($ ___ , ___ $)$

2. Sam and Matt divided a salary of $180 in a ratio of 4:5. Calculate how much each boy got.

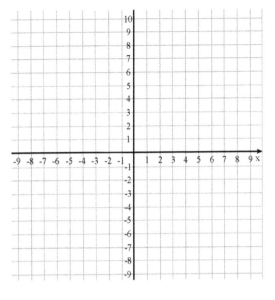

3. Find the greatest common factor of the given number pairs.

a. 56 and 70	b. 96 and 36

4. Find five numbers that are multiples of both 5 and 9.

5. Solve the equations by thinking logically.

a. $4 \times$ _____ $= 0.0012$	b. $0.2 \times$ _____ $= 0.06$	c. $0.03 \times$ _____ $= 30$

6. Solve the equations.

a. $0.5x = 30$	b. $0.01x = 2$	c. $c + 1.1097 = 3.29$

7. The grid represents a board game.
   Samantha has game pieces at (−50, 40) and (−50, −25).

   a. How far apart are Samantha's two game pieces
   from each other?

   b. Hailey guessed, "Your game piece is at (10, 40)."
   Samantha said, "You missed by _____ units!"

   c. Originally, Samantha had 6 game pieces in the game.
   What percentage of game pieces does she have left?

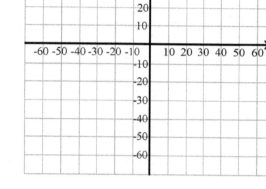

8. Find the better deal: an $18 flash drive is discounted by 15%,
   and another, $20 flash drive is discounted by 1/5.

9. Alice had a box of 90 oranges. She gave 3/5 of the oranges to Beatrice.
   Then, of what was left, she gave 1/4 to Michael.
   How many oranges does Alice have now?
   How many oranges did Michael get?

10. Asphalt paving costs $1,250 for 500 square feet. Fill in the equivalent rates.

$$\frac{\rule{2cm}{0.4pt}}{100 \text{ sq. ft.}} = \frac{\rule{2cm}{0.4pt}}{200 \text{ sq. ft.}} = \frac{\rule{2cm}{0.4pt}}{500 \text{ sq. ft.}} = \frac{\rule{2cm}{0.4pt}}{2{,}000 \text{ sq. ft.}} = \frac{\rule{2cm}{0.4pt}}{2{,}400 \text{ sq. ft.}}$$

11. Add and subtract.

a. $-2 + (-11) =$ _____	**b.** $-1 + (-7) =$ _____	**c.** $10 - 17 =$ _____	**d.** $7 - (-3) =$ _____
$(-11) + 2 =$ _____	$1 - 7 =$ _____	$-10 - 17 =$ _____	$-3 - (-7) =$ _____

12. Multiply, and shade the grid to illustrate the multiplications.

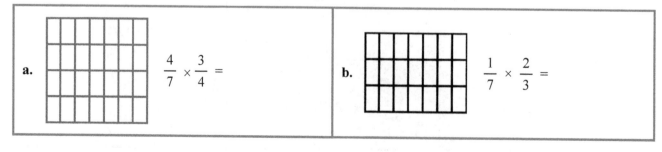

a. $\dfrac{4}{7} \times \dfrac{3}{4} =$

b. $\dfrac{1}{7} \times \dfrac{2}{3} =$

Cumulative Review, Grade 6, Chapters 1-10

A calculator is not allowed.

1. Find the perimeter and the area of this triangle.

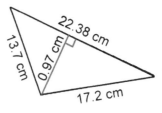

2. The distance from Ben's home to his workplace is only 0.7 miles.

   a. What is it in feet?

   b. Ben walks to work 4/5 of his workdays, and the rest of the time he rides a bike.
   Calculate how many miles he ends up walking in a year going to work.
   Assume that he works 48 weeks in a year, 5 days a week.

3. Convert the measurements into the given units.

	m	dm	cm	mm
a. 7.82 m	7.82			
b. 109 mm				109 mm

4. Divide, and give your answer as a decimal. If necessary, round the answers to three decimal digits.

a. $17.54 ÷ 3	**b.** 2.4 ÷ 0.05

5. **a.** What is the volume of a shoe box measuring 25 cm by 18 cm by 12 cm?

   b. Sketch the net of the shoe box.

   c. Calculate the surface area of the shoe box.

6. Write an expression.

   a. 5 less than x to the 5th power

   b. The quantity 2 minus x, cubed

   c. 2 times the sum of 10 and y

   d. The difference of s and 2, divided by s squared

7. Find the value of the expression in 6a above, if x has the value 2.

8. Factor these sums (writing them as products).

a. $56x + 14 =$ ____ (____ + ____)	**b.** $18u + 60 =$ ____ (____ + ____)

9. Solve the equations.

a. $\quad y \div 50 \;=\; 60 \cdot 2$	**b.** $\quad 3x - x \;=\; 3 + 7$	**c.** $\quad 7x \;=\; 50$
$=$	$=$	$=$
$=$	$=$	$=$
$=$	$=$	$=$

10. Solve the inequality $x - 12 > 6$ in the set $\{11, 13, 15, 17, 19, 21, 23\}$.

11. Mark the following numbers on this number line that starts at 0 and ends at 2.

$$0.3, \quad \frac{5}{4}, \quad 0.45, \quad 1.25, \quad \frac{3}{5}, \quad 1.07, \quad 1\frac{3}{10}, \quad 1.95, \quad \frac{1}{3}$$

12. Write these fractions as decimals. Give your answers to three decimal digits.

a. $\dfrac{5}{4} =$	b. $\dfrac{6}{7} =$	c. $\dfrac{19}{16} =$

13. A puzzle measures 8 1/2 inches by 10 inches. Calculate the area of the puzzle in square centimeters, using the fact that 1 inch = 2.54 cm.

14. Oats cost $0.92 per pound. Eric bought 2 3/4 lb.
    Calculate the total cost of Eric's purchase.

15. Make a stem-and-leaf plot from the following data:

Heart rates of a group of 13-year olds after doing jumping jacks for 30 seconds:

159   162   145   175   155   163   160   140   158   190   172   162   152   163   148   150

CPSIA information can be obtained
at www.ICGtesting.com
Printed in the USA
LVHW060212140821
695161LV00006B/290